WATER TREATMENT (PART I)

Municipal water treatment OR Water treatment for domestic and industrial purpose

Steps involved in municipal water treatment

1. Screening
2. Sedimentation
3. Coagulation
4. Filtration
5. Sterilization

1. Screening

The raw water is passed through screens, having large number of holes; here floating matters (for example hay, sticks, dead plants, etc) are retained by them.

2. Sedimentation

Sedimentation is the process of settle downing of suspended particles due to the action of gravity. Water is allowed to stand for a few days in a big sedimentation tanks where relatively bigger size suspended particles are settled to bottom. The clear water is then slowly removed without disturbing sediments. In order to remove floating impurities, screen of various kinds e.g. bar screen, band and drum screen are employed. Sedimentation tanks are of horizontal flow type and circular shaped upward flow type.

The process of sedimentation is influenced by (i) Horizontal flow of water, (ii) the size of the particle, (iii)the specific gravity of the particles and (iv) temperature of the water.

Sedimentation takes a long time, requires large capacity settling tanks and cannot be removed all coarse-dispersed matter.

3. Coagulation

Finely divided particles do not settle down in sedimentation process due to their small size. These particles are in colloidal form and are generally negatively charged and do not come together (coalesce) because of mutual repulsion. Such impurities are removed by adding chemicals of opposite charge. This chemical of opposite charge neutralize the charge on suspended collides and bring them together. Thus particles get aggregated and settle down to bottom. This process is called coagulation. The widely used coagulant is alum $\left[K_2(SO_4)_3 \cdot Al_2(SO_4)_3 \cdot 24H_2O\right]$.

For example $Al_2(SO_4)_3$ (common coagulant) when added in water, it forms $Al(OH)_3$

$$Al_2(SO_4)_3 + 6H_2O \longrightarrow 2Al(OH)_3 + 3H_2SO_4$$

which acts as a floc or coagulant. $Al(OH)_3$ has enormous surface area and removes the finely divided colloidal by neutralizing the charge and adsorption.

In water with no alkalinity Na_2CO_3 must be added so that effective flocculation can form

$$Al_2(SO_4)_3 + 3Na_2CO_3 + 3H_2O \longrightarrow 2Al(OH)_3 + 3Na_2SO_4 + 3CO_2$$

For treatment of acidic water, $NaAlO_2$ (sodium aluminate) is added a source of $Al(OH)_3$

$$NaAlO_2 + 2H_2O \longrightarrow NaOH + Al(OH)_3$$

The efficiency of the coagulation process can be increased by addition of lime, fuller's earth, bentonite clay and polyelectrolytes.

4. Filtration

Filtration is a process of removing coarse impurities from water by passing it through a porous material. Porous material used for segregation of course impurities is known as a filter. The common materials used for the filtering medium are quartz sand, containing SiO_2, crushed anthracite and porous clay.

Water for domestic use may be filtered through large area of finely graded sand beds at a slow rate $\left(9\,L\,ft^{-2}hr^{-1}\right)$.

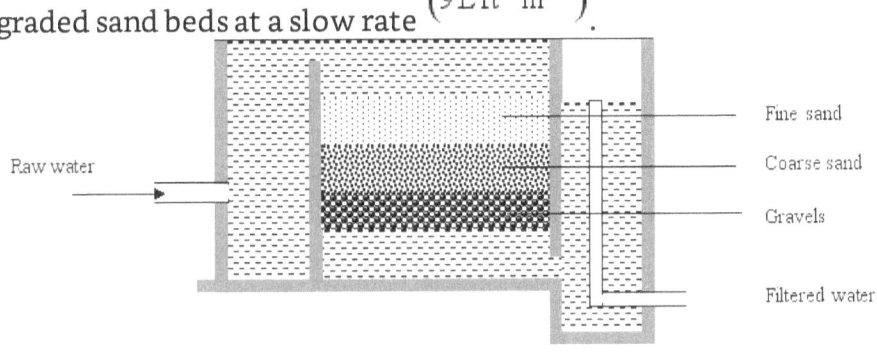

Fine sand

Raw water

Coarse sand

Gravels

Filtered water

Water for domestic use may be filtered through large area of finely graded sand beds at a slow rate. The rate of slowly diminishes due to the deposition of sediments in filter bed. This deposition is removed by scarping or washing of bed.

The rate of the filtration (rapid gravity filtration) can be achieved by using graded quartz sand. In this case sand bed is cleaned either by agitation with compressed air followed by flush of water current.

In industries rapid pressure filtration more widely used than gravity filter. Pressure filters are manufactured in vertical and horizontal types. In this case, the high rate of filtration is maintained by selecting two filtering materials (crushed anthracite & quartz) of different bulk mass and grain.

5. Sterilization of water

C hlorination

Removal microorganism (bacteria, protozoa, viruses, worms etc.) from water is known as sterilization. Chlorine is the most common sterilizing agent in water treatment. Chlorine may be added in the form of bleaching powder, or directly as gas or in the form of concentrated solution in water. Bleaching action in water given by

$$CaOCl_2 + H_2O \longrightarrow Ca(OH)_2 + Cl_2$$
$$Cl_2 + H_2O \longrightarrow HOCl + HCl$$
$$HOCl \longrightarrow [O] + HCl$$
$$HOCl \rightleftharpoons H^+ + OCl^-$$

(Hypochlorous acid) (Nascent Oxygen)

The $HOCl$ acid deactivates the enzymatic reaction in microorganisms. The death of micro-organism is due to this deactivation. The nascent oxygen so librated destroys the germs and bacteria by oxidation. The OCl^- ions are capable of rupturing the cell membrane of microbes. However, $HOCl$ is more disinfectant than reactive OCl^-. Thus the chlorine is effective disinfectant at lower pH.

Advantages of chlorine as disinfectant
(i)   It is stable and does not deteriorate on keeping.
(ii)   It can be used at high and low temperature.
(iii)   It introduces no salt impurities in treated water.

Disadvantages of chlorination
(i)   It is effective disinfectant at lower pH.
(ii)   Excess chlorine produced unpleasant taste and odour in the water.
(iii)   Excess chlorine cause irritation to the mucous membranes.

(iv)   The formation of trimethanes which is carcinogenic.

Excess chlorine can be removed by treatment with ammonia. Ammonia reacts with chlorine and forms tasteless compound chloramine (NH_2Cl).

Break-point chlorination

Sufficient chlorine is added to oxidize all the organic matter, destroyed bacteria and react with ammonia is known as break-point chlorination. This is practically realized by plotting residual chlorine versus actual chlorine dose added (shown in fig).

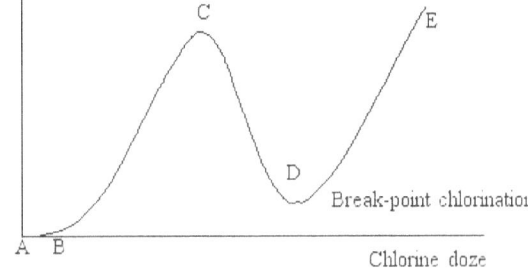

AB: This region corresponds to reaction of chlorine with ammonia to form monochloramine and dichloramine. Due to this residual chlorine is absent in this region.

$$Cl_2 + NH_3 \longrightarrow NH_2Cl + HCl$$
$$2Cl_2 + NH_3 \longrightarrow NHCl_2 + 2HCl$$

BC: This region shows linear increase in residual chlorine concentration with chlorine doze, indicates there is any notable oxidation of organic species.

CD: This segment of curve shows the fall in concentration of residual chlorine, this due to the fact that some organic compounds which defy oxidation at lower concentration of chlorine get oxidized.

The point 'D' corresponds to breakpoint chlorination, beyond it the residual chlorine appearing more or less agrees with added chlorine.

Advantages (significance) of Break point chlorination

1. It completely removes the ammonia, micro-organism and organic matter.
2. This method removes unpleasant odour and taste from the water.
3. It prevents the growth of any weeds in water.
4. It destroyed all the disease-producing bacterias.

Impurities in water

Inorganic impurities

(i)   Cations: $Ca^{2+}, Mg^{2+}, Fe^{2+}, Al^{3+}, K^+$ and Na^+

(ii)   Anions: $OH^-, Cl^-, SO_4^{2-}, NO_3^-, HCO_3^-$ and CO_3^{2-}

Clay and sands

Dissolved gases: CO_2, O_2, N_2, H_2S and NH_3

Organic impurities: Oil, vegetable, aminoacids, surfactants, coloring matter etc.

Hardness of water

Hardness in water is due to presence of salts of calcium, magnesium and other heavy metals. Sample hard water, when treated with soap, it forms insoluble scum which do not possess any detergent value. For example

$$C_{17}H_{35}COONa + Ca^{2+} \longrightarrow (C_{17}H_{35}COO)_2 Ca \downarrow + 2NaCl$$
$$(Scum)$$

Due to this interaction of soap with hardness producing substances, does not produce lather. Thus hardness defined as the soap consuming ability of water.

Types of hardness

1. **Temporary hardness or Carbonate hardness**:
Temporary hardness in water is caused by the presence of dissolved bicarbonates calcium, magnesium and other heavy metals. Temporary hardness is removed by boiling water

$$Ca(HCO_3)_2 \xrightarrow{\Delta} CaCO_3\downarrow + H_2O + CO_2\uparrow$$
$$Mg(HCO_3)_2 \xrightarrow{\Delta} Mg(OH)_2\downarrow + 2CO_2\uparrow$$

2. **Permanent hardness or non carbonate hardness**:
Permanent hardness is due to presence of chloride, sulphate and nitrate of calcium, magnesium and other heavy metals. Permanent hardness cannot be removed by boiling.

Units of hardness:

1. Parts per million (ppm):
Parts per million is the parts of calcium carbonate equivalent hardness per 10^6 parts of water.

2. Milligrams per litre (mg/L):
Milligrams per litre is the number of milligrams of $CaCO_3$ equivalent hardness present per litre of water.

3. Clarke's degree $(^{\circ}Cl)$:
Clarke's degree is the parts of $CaCO_3$ equivalent hardness per 7×10^4 part of water.

4. Degree French:
Degree French is the parts of $CaCO_3$ equivalent hardness per 10^5 parts of water

Thus inter-relationship between various units
$$1ppm = 1mg/L = 0.1^{\circ}Fr = 0.07^{\circ}Cl$$

EQUIVALENT OF $CaCO_3$

The concentration of all causing hardness should be expressed in equivalent of $CaCO_3$. The molar mass of $CaCO_3$ is 100 which make convenience in conversion, however, $CaCO_3$ is most insoluble salt. Thus,

$$\text{Equivalent of } CaCO_3 = \frac{[\text{Molar Mass of } CaCO_3]}{[\text{Molar Mass of impurity}]} \times [\text{Mass of impurity}]$$

Conversion Table

Dissolved salt	Molar Mass	Conversion factor
$CaCO_3$	100	100/100
$Ca(HCO_3)_2$	162	100/162
$Mg(HCO_3)_2$	146	100/146
$CaSO_4$	136	100/136
$CaCl_2$	111	100/111
$MgSO_4$	120	100/120
$MgCl_2$	95	100/95
$MgCO_3$	84	100/84
$Mg(NO_3)_2$	148	100/148
Ca^{2+}	40	100/40
Mg^{2+}	24	100/24
CO_2	44	100/44

HCO_3^-	61	100/122
$NaHCO_3^-$	84	100/168
OH^-	17	100/34
H_2SO_4	98	100/98
HCl	36.5	100/73
H^+	1	100/2
CO_3^{2-}	60	100/60
$NaAlO_2$	82	100/164
$Al_2(SO_4)_3$	342	100/114
$FeSO_4 \cdot 7H_2O$	278	100/278

Note: Substances which do not contribute towards hardness are KCl, $NaCl$, SiO_2, Na_2SO_4, Fe_2O_3, K_2SO_4 and ignore while calculating hardness.

All carbonates and bicarbonates of calcium and magnesium are taken as temporary hardness.

All chlorides, sulphates and nitrates of calcium and magnesium are taken as permanent hardness.

Example

1. Calculate temporary, permanent and total hardness of a sample of water conataining:

$Mg(HCO_3)_2 = 7.3 mg/L$, $\quad Ca(HCO_3)_2 = 16.2 mg/L$, $\quad MgCl_2 = 9.5 mg/L$, $CaSO_4 = 13.6 mg/L$, $\quad SiO_2 = 10 mg/L$, $\quad NaCl = 7 mg/L$

Solu.

Impurity	Concentration (mg/L)	$CaCO_3$ equivalent (mg/L)
$Mg(HCO_3)_2$	7.3	$(100/146) \times 7.3 = 5$
$Ca(HCO_3)_2$	16.2	$(100/162) \times 16.2 = 10$

$MgCl_2$	9.5	$(100/95) \times 9.5 = 10$
$CaSO_4$	13.6	$(100/136) \times 13.6 = 10$

Temporary hardness $= [5+10] = 15 \text{mg/L}$

Permanent hardness $= [10+10] = 20 \text{mg/L}$

Total hardness $= [15+20] = 35 \text{mg/L}$

2. A sample of water contains the following dissolved salts:

$Mg(HCO_3)_2 = 22 \text{mg/L}$, $Ca(HCO_3)_2 = 16.2 \text{mg/L}$, $MgCl_2 = 30 \text{mg/L}$,

$CaCl_2 = 85 \text{mg/L}$, $CaSO_4 = 28 \text{mg/L}$, $Fe_2O_3 = 10 \text{mg/L}$, $NaCl = 11 \text{mg/L}$,

$Mg(NO_3)_2 = 29.6 \text{mg/L}$, $CaCO_3 = 20 \text{mg/L}$,

Calculate temporary and permanent hardness. Express results in ppm, $^{\circ}Cl$ and $^{\circ}Fr$

EDTA method for determination of hardness

1. Standard water contains 1.5g of $CaCO_3$ per liter 20 ml of this required 25 ml of EDTA solution. 100 ml of water sample required 18 ml EDTA solution. The same sample of the boiling required 12 ml EDTA solution. Calculate the temporary hardness of the given sample of water, in terms of ppm. $CaCO_3$ $CaCO_3$ $CaCO_3$

Solu.: Molarity of standard solution

1.5 g of $CaCO_3$ is dissolved in 1 Lit. distilled water

$$\text{Molarity}(M_1) = \frac{W(g) \times 1000}{M \times V \,(\text{ml of solution})}$$

$$\text{Molarity}(M_1) = \frac{1.5 \times 1000}{100 \times 1000} = 0.015\,M$$

Molarity of EDTA solution

V_1 (Std. solution) = 20 ml, M_1 (Std. solution) = 0.015 M

V_2 (EDTA solu.) = 25 ml, M_2 (EDTA solu.) = ?

(Std. solu.) $M_1 V_1 = M_2 V_2$ (EDTA solu.), $M_2 = \dfrac{0.015 \times 20}{25} = 0.012 M$

Determination of total hardness

V_2' (EDTA solu.) = 18 ml, $M_2 = 0.012 M$

V_3 (Water sample) = 100 ml, M_3 (Water sample) = ?

$M_3 = \dfrac{0.012 \times 18}{100} = 0.00216 M$

Strength   = Molarity × Mol. wt.

= 0.00216 × 100 g/L = 0.216 g/L

Total hardness = 0.216 × 1000 mg/L

= 216 mg/L or ppm

Determination of permanent hardness

V_2' (EDTA solu.) = 12 ml, $M_2 = 0.012 M$

V_3 (Water sample) = 100 ml, M_3' (Water sample) = ?

$M_3' = \dfrac{0.012 \times 12}{100} = 0.00144 M$

Strength   = Molarity × Mol. wt.

= 0.00144 × 100 g/L = 0.144 g/L

Permanent hardness   = 0.144 × 1000 mg/L

= 144 mg/L or ppm

Temporary hardness = Total hardness – Permanent hardness

Temporary hardness = 216 – 144 = 72 ppm

2. Calculate the hardness of a water sample, whose 10 ml required 20 ml of EDTA. 20 ml of $CaCl_2$ solution whose strength is equivalent 1.5 g of $CaCO_3$ liter, required 30 ml of EDTA solution.

Solution

Determination of Molarity of standard solution,

1.5 g of $CaCO_3$ is dissolved in 1 Lit. distilled water

$\text{Molarity} (M_1) = \dfrac{W (g) \times 1000}{M \times V \text{ (ml of solution)}}$

$$\text{Molarity}\,(M_1) = \frac{1.5 \times 1000}{100 \times 1000} = 0.015\,M$$

Determination of Molarity of EDTA solution

V_1 (Std. solution) = 20 ml, M_1 (Std. solution) = 0.015 M

V_2 (EDTA solu.) = 30 ml, M_2 (EDTA solu.) = ?

(Std. solu.) $M_1 V_1 = M_2 V_2$ (EDTA solu.), $M_2 = \dfrac{0.015 \times 20}{30} = 0.01M$

Determination of total hardness

V_2' (EDTA solu.) = 20 ml, $M_2 = 0.01M$

V_3 (Water sample) = 10 ml, M_3 (Water sample) = ?

$M_3 = \dfrac{0.01 \times 20}{10} = 0.02M$

Total hardness $= 0.02 \times 100 \times 1000\,mg/L$

$\qquad\qquad = 2000\,mg/L$ or ppm

3. 50 ml of a standard hard water containing 1 mg of pure $CaCO_3$ per ml consumed 20 ml of EDTA. 50ml of a water sample consumed 25ml of the same EDTA solution using EBT. Calculate the total hardness of water sample in ppm

Hint:

Molarity of standard solution

1 mg of $CaCO_3$ is dissolved in 1ml distilled water. It means 1000mg (1g) of $CaCO_3$ is dissolved in 1000ml distilled water $CaCO_3$. Calculate the molarity of standard solution.

WATER SOFTENING

The processes of removing hardness-producing salts from water, is known as softening of water. Lime-soda process is the most important chemical for water softening.

The principle: Chemically convert all the soluble hardness causing impurities into insoluble precipitates by adding calculated amount of lime $\left[Ca\left(OH\right)_2\right]$ and soda $\left[Na_2CO_3\right]$. Insoluble precipitates are then removed by settling and filtration.

Lime-soda process

In this method, calculated quantity of lime and soda are mixed with water at room temperature. At this temperature the ppt formed are finely divided, which can be removed using appropriate coagulant. Cold lime-soda process provides water, containing a hardness of 50 to 60 ppm. The various reactions involved in this process are given below:

Action of lime

1. Lime removes temporary hardness

$$Ca\left(HCO_3\right) + Ca\left(OH\right)_2 \longrightarrow 2CaCO_3 \downarrow + 2H_2O$$

$$Mg\left(HCO_3\right)_2 + 2Ca\left(OH\right)_2 \longrightarrow Mg\left(OH\right)_2 \downarrow + 2CaCO_3 \downarrow + 2H_2O$$

2. Lime removes the permanent magnesium hardness

$$MgSO_4 + Ca\left(OH\right)_2 \longrightarrow Mg\left(OH\right)_2 \downarrow + CaSO_4$$

$$MgCl_2 + Ca\left(OH\right)_2 \longrightarrow Mg\left(OH\right)_2 \downarrow + CaCl_2$$

$$Mg\left(NO_3\right)_2 + Ca\left(OH\right)_2 \longrightarrow Mg\left(OH\right)_2 \downarrow + Ca\left(NO_3\right)_2$$

3. Lime removes the iron and aluminium salts

$$FeSO_4 + Ca\left(OH\right)_2 \longrightarrow Fe\left(OH\right)_2 \downarrow + CaSO_4$$

$$Al_2\left(SO_4\right)_3 + 3Ca\left(OH\right)_2 \longrightarrow 2Al\left(OH\right)_3 \downarrow + 3CaSO_4$$

4. Lime remove free mineral acids

$$2HCl + Ca(OH)_2 \longrightarrow CaCl_2 + 2H_2O$$
$$H_2SO_4 + Ca(OH)_2 \longrightarrow CaSO_4 + 2H_2O$$

5. Lime removes dissolved CO_2 and H_2S

$$CO_2 + Ca(OH)_2 \longrightarrow CaCO_3 \downarrow + H_2O$$
$$H_2S + Ca(OH)_2 \longrightarrow CaS \downarrow + 2H_2O$$

6. Soda removes permanent hardness

$$CaCl_2 + Na_2CO_3 \longrightarrow CaCO_3 \downarrow + 2NaCl$$
$$CaSO_4 + Na_2CO_3 \longrightarrow CaCO_3 \downarrow + Na_2SO_4$$
$$Ca(NO_3)_2 + Na_2CO_3 \longrightarrow CaCO_3 \downarrow + 2NaNO_3$$

Cold Lime-Soda Process and its type

(i)   The intermittent (batch process) type,
(ii)   The conventional type
(iii)   The catalyst or spiractor type, and
(iv)   The sludge blanket type

1. The intermittent (batch process) type

The intermittent type of lime-soda softener consists of a set of two tanks which are used in turn. Raw water and calculated quantities of the chemicals are slowly sent into the tank simultaneously under agitation with the help of the stirrer. Some sludge from a previous operation is also added which forms nucleus for fresh precipitation and thus accelerate the process. After certain interaction time, stirring is stopped and allow sludge to settle down. The clear water is then removed and sent for further process.

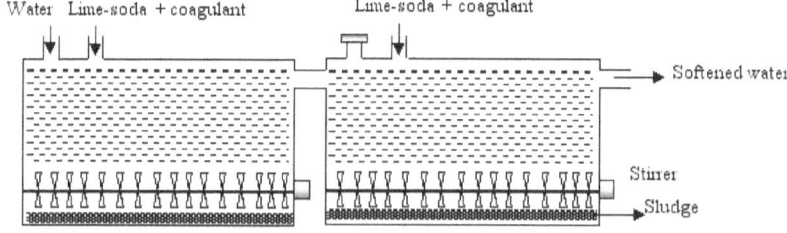

2. **Conventional type**

In conventional type lime soda process, the raw water and chemicals (lime-soda-coagulant) fed from the top into the vertical chamber where stirrer is mounted.

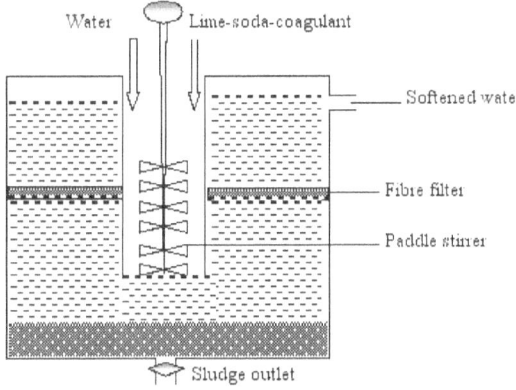

Due to continuous stirring raw water and chemical mixed thoroughly. The sludge is periodically removed through the bottom. The soften water rising up and passes through the fibre-filter where traces of sludge are removed.

3.   **Catalyst or Spiractor type**:

The spiractor consists of a conical tank which is two-thirds filled with finely divided granular catalyst. The catalyst used here is insoluble mineral such as calcite or sand. The calculate quantity of lime-soda and raw water send through opening situated at the bottom of the tank. Water moves upward under pressure. The sludge form in this process is deposited on surface of the catalyst.

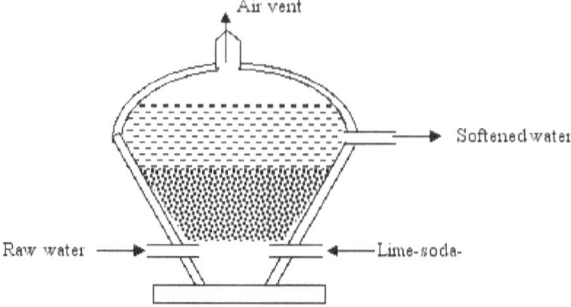

4. Sludge-blanket type lime-soda process

In this process coagulated water is not dropping down but filters it upwardly through a suspended sludge blanket. This prolonged intimate contact with the floc help in fully utilization of coagulant or other absorbents such as activated charcoal, lime and soda. In this single unit all three processes viz mixing, softening and clarification.

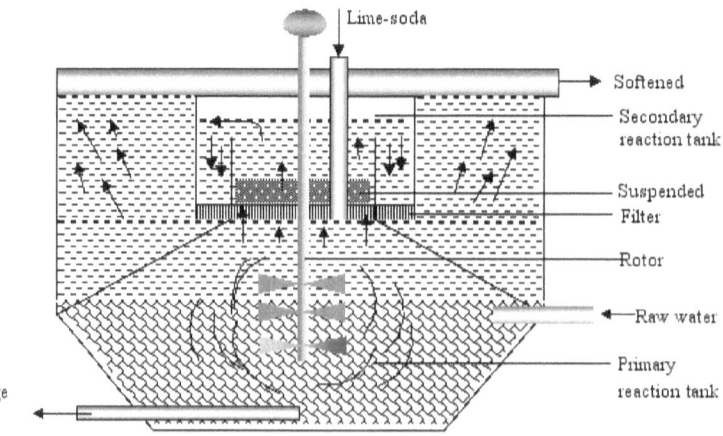

Features of Sludge-blanket type lime-soda process

(i)   Intimate contact between water and chemicals ensures complete utilization of lime-soda.

(ii)   Intimate contact with a large mass of solid phase mostly prevents supersaturation.

(iii)   Retention period is small as compare to conventional lime-soda method.

(iv)   Due to higher efficiency and lesser turbid, this method is employed for many industrial applications.

Hot Lime-soda process

Hot lime-soda process involves in treating water with softening chemicals at a temperature of 94-100 ^{0}C.

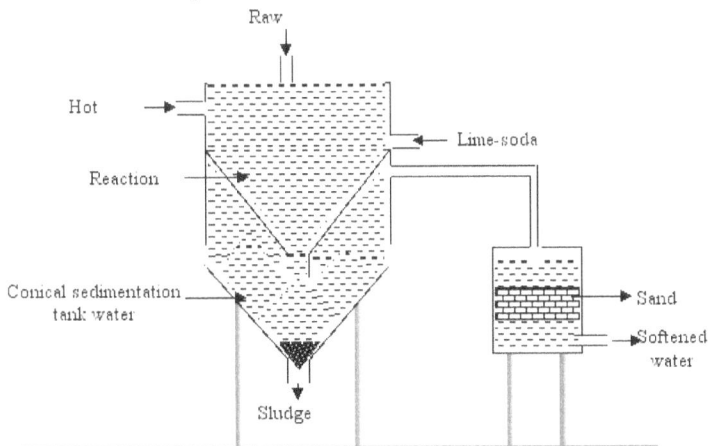

The main features of this process are

(i)   The rate softening is high.

(ii)   The precipitation reaction proceeds with faster rate.

(iii)   Fewer amounts of chemicals are required.

(iv)   Rate of nucleation also increased.

(v)   No coagulation is needed, as sludge is settled down rapidly.

(vi)   Due to high temperature some dissolved gases are driven out to some extent.

The hardness of water 17-34 ppm.

Calculation of lime and soda

Constituent	Reactions	Lime/soda
$Ca(HCO_3)_2$	$Ca(HCO_3)_2 + Ca(OH)_2 \longrightarrow 2CaCO_3 + 2H_2O$	L
$Mg(HCO_3)_2$	$Mg(HCO_3)_2 + 2Ca(OH)_2 \longrightarrow 2CaCO_3 + Mg(OH)_2 + 2H_2O$	2 L
Ca^{2+}	$Ca^{2+} + Na_2CO_3 \longrightarrow CaCO_3 + 2Na^+$	S
Mg^{2+}	$Mg^{2+} + Ca(OH)_2 \longrightarrow Ca^{2+} + Mg(OH)_2$ $Ca^{2+} + Na_2CO_3 \longrightarrow CaCO_3 + 2Na^+$	L + S
HCO_3^-	$2HCO_3^- + Ca(OH)_2 \longrightarrow CaCO_3 + H_2O + CO_3^{2-}$	L − S
H^+	$2H^+ + Ca(OH)_2 \longrightarrow Ca^{2+} + 2H_2O$ $Ca^{2+} + Na_2CO_3 \longrightarrow CaCO_3 + 2Na^+$	L + S

CO_2	$CO_2 + Ca(OH)_2 \longrightarrow CaCO_3 + H_2O$	L
Fe^{2+}	$Fe^{2+} + Ca(OH)_2 \longrightarrow Ca^{2+} + Fe(OH)_2$ $Ca^{2+} + Na_2CO_3 \longrightarrow CaCO_3 + 2Na^+$	L+S
Al^{3+}	$2Al^{3+} + 3Ca(OH)_2 \longrightarrow 3Ca^{2+} + 2Al(OH)_3$ $3Ca^{2+} + 2Na_2CO_3 \longrightarrow 3CaCO_3 + 6Na^+$	L+S
treated water contains OH^-	Presence of OH^- in treated water is assumed to be supplied by equivalent amount of $Ca(OH)_2$. But, however, it adds Ca^{2+} in water which should have been removed by equivalent amount of soda.	L+S
treated water contains CO_3^{2-}	The CO_3^{2+} in treated water are required to be present must have been supplied by its equivalent amount of Na_2CO_3.	S
$NaAlO_2$	$NaAlO_2 + H_2O \longrightarrow Al(OH)_3 + NaOH$ Presence of OH^- assume to be supplied by $Ca(OH)_2$ and hence the corresponding quantity of $NaAlO_2$ should be deducted both from lime as well as soda requirements.	$-L-S$

Calculation of lime and soda

Constituent	Reactions	Lime/soda
$Ca(HCO_3)_2$	$Ca(HCO_3)_2 + Ca(OH)_2 \longrightarrow 2CaCO_3 + 2H_2O$	L
$Mg(HCO_3)_2$	$Mg(HCO_3)_2 + 2Ca(OH)_2 \longrightarrow 2CaCO_3 + Mg(OH)_2 + 2H_2O$	2 L
CO_2	$CO_2 + Ca(OH)_2 \longrightarrow CaCO_3 + H_2O$	L
Mg^{2+}	$Mg^{2+} + Ca(OH)_2 \longrightarrow Ca^{2+} + Mg(OH)_2$ $Ca^{2+} + Na_2CO_3 \longrightarrow CaCO_3 + 2Na^+$	L+S

H^+	$2H^+ + Ca(OH)_2 \longrightarrow Ca^{2+} + 2H_2O$ $Ca^{2+} + Na_2CO_3 \longrightarrow CaCO_3 + 2Na^+$	$L+S$
Fe^{2+}	$Fe^{2+} + Ca(OH)_2 \longrightarrow Ca^{2+} + Fe(OH)_2$ $Ca^{2+} + Na_2CO_3 \longrightarrow CaCO_3 + 2Na^+$	$L+S$
Al^{3+}	$2Al^{3+} + 3Ca(OH)_2 \longrightarrow 3Ca^{2+} + 2Al(OH)_3$ $3Ca^{2+} + 2Na_2CO_3 \longrightarrow 3CaCO_3 + 6Na^+$	$L+S$
treated water contains OH^-	Presence of OH^- in treated water is assumed to be supplied by equivalent amount of $Ca(OH)_2$. But, however, it adds Ca^{2+} in water which should have been removed by equivalent amount of soda.	$L+S$
Ca^{2+}	$Ca^{2+} + Na_2CO_3 \longrightarrow CaCO_3 + 2Na^+$	S
treated water contains CO_3^{2-}	The CO_3^{2+} in treated water are required to be present must have been supplied by its equivalent amount of Na_2CO_3.	S
HCO_3^-	$2HCO_3^- + Ca(OH)_2 \longrightarrow CaCO_3 + H_2O + CO_3^{2-}$	$L-S$
$NaAlO_2$	$NaAlO_2 + H_2O \longrightarrow Al(OH)_3 + NaOH$ Presence of OH^- assume to be supplied by $Ca(OH)_2$ and hence the corresponding quantity of $NaAlO_2$ should be deducted both from lime as well as soda requirements.	$-L-S$

Types of hardness / dissolved impurity	Constituents	Lime	soda
Temporary Ca hardness	$Ca(HCO_3)_2$, $CaCO_3$	L	
Temporary Mg hardness	$Mg(HCO_3)_2$, $MgCO_3$	$2L$	
Permanent Ca	Ca^{2+}, $CaSO_4$, $CaCl_2$, $Ca(NO_3)_2$		S

hardness			
Permanent Mg hardness	Mg^{2+}, $MgSO_4$, $MgCl_2$, $Mg(NO_3)_2$	L	S
Fe^{2+} impurity	$FeSO_4 \cdot 7H_2O$	L	S
Al^{3+} impurity	$Al_2(SO_4)_3$	L	S
H^+	HCl, H_2SO_4, HNO_3, any acid which furnishes H^+	L	S
OH^-	Present in treated water	L	S
CO_3^{2-}	Present in treated water		S
Dissolved CO_2		L	
Bicarbonates (HCO_3^-)	HCO_3^-, $NaHCO_3^-$	L	-S
$NaAlO_2$	Coagulant	-L	-S

Note: Substances which do not contribute towards hardness are KCl, $NaCl$, SiO_2, Na_2SO_4, Fe_2O_3, K_2SO_4 and ignore while calculating lime and soda.

Let V be the volume of water which to be softened and x be the percentage purity of commercial lime and soda available in the market, then

$$Total\,requirement\,of\,Lime = \frac{74}{100}\begin{cases} Temporary\,Ca\,hardness + 2\times Temp.\,Mg\,hardness \\ +Permenent\,Mg\,hardness + dissolved\,CO_2 + Fe^{2+} \\ +Al^{3+} + HCl + H_2SO_4 + HCO_3^- - NaAlO_2 \\ +OH^-\,from\,treated\,water \end{cases} \times \frac{100}{x} \times \frac{V}{10^6}kg$$

$$Total\,requirement\,of\,Soda = \frac{106}{100}\begin{cases} Permanent\,Ca\,hardness + Permenent\,Mg\,hardness \\ +Fe^{2+} + Al^{3+} + H_2SO_4 + HCl - HCO_3^- - NaAlO_2 \\ +\left(OH^- + CO_3^{2-}\right)\,from\,treated\,water \end{cases} \times \frac{100}{x} \times \frac{V}{10^6}kg$$

Examples

1. Calculate the amount of lime and soda for the treatment of 1 million litres of water containing:

$Mg(HCO_3)_2 = 14.6\,mg/L$, $\quad$ $Ca(HCO_3)_2 = 8.1\,mg/L$, $\quad$ $MgCl_2 = 38\,mg/L$,

$CaCl_2 = 33.3\,mg/L$, $\quad$ $HCO_3^- = 91.5\,mg/L$.

The coagulant $Al_2(SO_4)_3$ was added at the rate of $17.1 mg/L$ of water.

Solu:

Impurity	$CaCO_3$ equivalent	Lime	Soda
$Mg(HCO_3)_2 = 14.6 mg/L$	$\dfrac{100}{146} \times 14.6 = 10 mg/L$	20	
$Ca(HCO_3)_2 = 8.1 mg/L$	$\dfrac{100}{162} \times 8.1 = 5 mg/L$	5	
$MgCl_2 = 38 mg/L$	$\dfrac{100}{95} \times 38 = 40 mg/L$	40	40
$CaCl_2 = 33.3 mg/L$	$\dfrac{100}{111} \times 33.3 = 30 mg/L$		30
$HCO_3^- = 91.5 mg/L$	$\dfrac{100}{122} \times 91.5 = 75 mg/L$	75	-75
$Al_2(SO_4)_3 = 17.1 mg/L$	$\dfrac{100}{114} \times 17.1 = 15 mg/L$	15	15
	Total	155	10

$$Total\ requirement\ of\ Lime = \frac{74}{100} \left\{ \begin{array}{l} Temporary\ Ca\ hardness + 2 \times Temp.\ Mg\ hardness \\ +Permenent\ Mg\ hardness + dissolved\ CO_2 + Fe^{2+} \\ +Al^{3+} + HCl + H_2SO_4 + HCO_3^- - NaAlO_2 \\ +OH^-\ from\ treated\ water \end{array} \right\} \times \frac{100}{x} \times \frac{V}{10^6} kg$$

$$Total\ requirement\ of\ Lime = \frac{74}{100} \{5 + 2 \times 10 + 40 + 15 + 75\} \times \frac{10^6}{10^6} kg$$

$$= \frac{74}{100} \times 155 kg = 114.7 kg$$

$$Total\ requirement\ of\ Soda = \frac{106}{100} \left\{ \begin{array}{l} Permanent\ Ca\ hardness + Permenent\ Mg\ hardness \\ +Fe^{2+} + Al^{3+} + H_2SO_4 + HCl - HCO_3^- - NaAlO_2 \\ +\left(OH^- + CO_3^{2-}\right)\ from\ treated\ water \end{array} \right\} \times \frac{100}{x} \times \frac{V}{10^6} kg$$

$$\text{Total requirement of Soda} \ = \frac{106}{100} \{30 + 40 + 15 - 75\} \times \frac{10^6}{10^6} \text{kg}$$

$$= \frac{106}{100} \times 10 \text{kg} = 10.6 \text{kg}$$

2. A water sample gave the following constituents on analysis:

$Mg(HCO_3)_2 = 73 \text{mg/L}$, $\quad Ca(HCO_3)_2 = 81 \text{mg/L}$, $\quad MgCl_2 = 95 \text{mg/L}$,
$CaSO_4 = 68 \text{mg/L}$, $\quad MgSO_4 = 12 \text{mg/L}$, $\quad NaCl = 4.5 \text{mg/L}$.

Calculate the cost of the chemicals required for softening 20,000 L of water, if the purity of lime is 95 % and that of soda is 90 %. The costs per 100 kg each of lime and soda are Rs. 75 and Rs. 2,480 respectively.

Solu.:

Impurity	$CaCO_3$ equivalent	Lime	Soda
$Mg(HCO_3)_2 = 73 \text{mg/L}$	$\frac{100}{146} \times 73 = 50 \text{mg/L}$	100	
$Ca(HCO_3)_2 = 81 \text{mg/L}$	$\frac{100}{162} \times 81 = 50 \text{mg/L}$	50	
$MgCl_2 = 95 \text{mg/L}$	$\frac{100}{95} \times 95 = 100 \text{mg/L}$	100	100
$CaSO_4 = 68 \text{mg/L}$	$\frac{100}{136} \times 68 = 50 \text{mg/L}$		50
$MgSO_4 = 12 \text{mg/L}$	$\frac{100}{120} \times 12 = 10 \text{mg/L}$	10	10
$NaCl = 4.5 \text{mg/L}$	Ignored		
	Total	260	160

$$\text{Total requirement of Lime} = \frac{74}{100} \left\{ \begin{array}{l} \text{Temporary Ca hardness} + 2 \times \text{Temp. Mg hardness} \\ + \text{Permenent Mg hardness} + \text{dissolved } CO_2 + Fe^{2+} \\ + Al^{3+} + HCl + H_2SO_4 + HCO_3^- - NaAlO_2 \\ + OH^- \text{ from treated water} \end{array} \right\} \times \frac{100}{x} \times \frac{V}{10^6} \text{kg}$$

where x and V are the percentage purity and volume of water respectively.

$$\text{Total requirement of Lime} = \frac{74}{100}\{50 + 2 \times 50 + (100 + 10)\} \times \frac{100}{95} \times \frac{20,000}{10^6} \text{kg}$$

$$= \frac{74}{100}\{260\} \times \frac{100}{95} \times \frac{20,000}{10^6} \text{kg} = 4.05\text{kg}$$

$$\text{Total requirement of Soda} = \frac{106}{100}\left\{\begin{array}{l}\text{Permanent Ca hardness} + \text{Permenent Mg hardness} \\ + Fe^{2+} + Al^{3+} + H_2SO_4 + HCl - HCO_3^- - NaAlO_2 \\ + \left(OH^- + CO_3^{2-}\right) \text{from treated water}\end{array}\right\} \times \frac{100}{x} \times \frac{V}{10^6}\text{kg}$$

$$\text{Total requirement of Soda} = \frac{106}{100}\{100 + 50 + 10\} \times \frac{100}{90} \times \frac{20,000}{10^6}\text{kg}$$

$$= \frac{106}{100}\{160\} \times \frac{100}{90} \times \frac{20,000}{10^6}\text{kg} = 3.76\text{kg}$$

The cost of lime is Rs. 75 per 100 kg,

$$\therefore \text{Total cost of lime} = \text{Rs. } 4.05 \times \frac{75}{100} = \text{Rs. } 3.03$$

The cost of soda is Rs. 2480 per 100 kg,

$$\therefore \text{Total cost of Soda} = \text{Rs. } 3.76 \times \frac{2480}{100} = \text{Rs. } 93.25$$

Total cost of Lime-soda treatment is Rs. 96.3

3. A water-works has to supply $1 \text{ m}^3/\text{s}$ of water. Raw water contain:

$Mg(HCO_3)_2 = 219$ppm, $Mg^{2+} = 36$ppm, $HCO_3^- = 18.3$ppm, $H^+ = 1.5$ppm

Calculate the cost of treating water per day, if lime (90% pure) and soda (955 pure) cost Rs. 500 per tone and Rs. 7,000 per tonne respectively.

Sol:

Impurity	$CaCO_3$ equivalent	Lime	Soda
$Mg(HCO_3)_2 = 219$ppm	$\frac{100}{146} \times 219 = 150$ppm	300	

$Mg^{2+} = 36ppm$	$\dfrac{100}{24} \times 36 = 150ppm$	150	150
$HCO_3^- = 18.3ppm$	$\dfrac{100}{122} \times 18.3 = 15ppm$	15	-15
$H^+ = 1.5ppm$	$\dfrac{100}{2} \times 1.5 = 75ppm$	75	75
	Total	540	210

Total volume of water supply per day

$$1m^3 = (10\,dm)^3 = 1000dm^3 = 1000L$$
$$1day = 24\,hrs. = 24 \times 60\,min. = 24 \times 60 \times 60\,sec = 86400\,sec$$
$$1sec = \frac{1}{24 \times 60 \times 60}\,day = \frac{1}{86400}\,day$$
$$V = 1000 \times 60 \times 60 \times 24\,L = 86,40,000L$$
$$1tonne = 1000\,kg$$

$$Total\,req.\,of\,Lime = \frac{74}{100}\begin{Bmatrix} Temporary\ Ca\ hardness + 2 \times Temp.\ Mg\ hardness \\ +Permenent\ Mg\ hardness + dissolved\ CO_2 + Fe^{2+} \\ +Al^{3+} + HCl + H_2SO_4 + HCO_3^- - NaAlO_2 \\ +OH^-\ from\ treated\ water \end{Bmatrix} \times \frac{100}{x} \times \frac{V}{10^6}kg$$

$$Total\,req.\,of\,Lime = \frac{74}{100}\{2 \times 150 + 150 + 15 + 75\} \times \frac{100}{90} \times \frac{86,40,000}{10^6}kg$$

$$= \frac{74}{100}\{540\} \times \frac{100}{90} \times \frac{86,40,000}{10^6}kg = 3836.16kg$$

$$Total\,req.\,of\,Soda = \frac{106}{100}\begin{Bmatrix} Permanent\ Ca\ hardness + Permenent\ Mg\ hardness \\ +Fe^{2+} + Al^{3+} + H_2SO_4 + HCl - HCO_3^- - NaAlO_2 \\ +\left(OH^- + CO_3^{2-}\right) from\ treated\ water \end{Bmatrix} \times \frac{100}{x} \times \frac{V}{10^6}kg$$

$$Total\,req.\,of\,Soda = \frac{106}{100}\{150 - 15 + 75\} \times \frac{100}{95} \times \frac{86,40,000}{10^6}kg$$

$$= \frac{106}{100}\{210\} \times \frac{100}{95} \times \frac{86,40,000}{10^6}kg = 2024.48kg$$

The cost of lime is Rs. 500 per tonne (1000 kg),

$$\therefore \text{Total cost of lime} = \text{Rs. } 3836.16 \times \frac{500}{1000} = \text{Rs. } 1918.08$$

The cost of soda is Rs. 7000 per tonne (1000 kg),

$$\therefore \text{Total cost of Soda} = \text{Rs. } 2024.48 \times \frac{7000}{1000} = \text{Rs. } 14171.36$$

Total cost of Lime-soda treatment = Rs. (1918.08 + 14171.36) = Rs. 1,60,089.45

4.   The water contains

$Ca^{2+} = 160ppm$, $Mg^{2+} = 72ppm$, $HCO_3^- = 732ppm$, Dissolved $CO_2 = 44ppm$

Calculate the quantities of lime and soda required for cold softening of 2, 00,000 L of water using 16.4 ppm of $NaAlO_2$ as a coagulant. On analysis the treated water contains:

$OH^- = 17ppm$, $CO_3^{2-} = 30ppm$

Sol.

Impurity	$CaCO_3$ equivalent	Lime	Soda
$Ca^{2+} = 160ppm$	$\frac{100}{40} \times 160 = 400ppm$		400
$Mg^{2+} = 72ppm$	$\frac{100}{24} \times 72 = 300ppm$	300	300
$HCO_3^- = 732ppm$	$\frac{100}{122} \times 732 = 600ppm$	600	-600
$CO_2 = 44ppm$	$\frac{100}{44} \times 44 = 100ppm$	100	
$NaAlO_2 = 16.4ppm$	$\frac{100}{164} \times 16.4 = 10ppm$	-10	-10

$OH^- = 17ppm$	$\dfrac{100}{34} \times 17 = 50ppm$	50	50
$CO_3^{2-} = 30ppm$	$\dfrac{100}{60} \times 30 = 50ppm$		50
	Total	1040	190

$$\text{Total req. of Lime} = \frac{74}{100}\{300+600+100-10+50\}\times\frac{2,00,000}{10^6}kg$$

$$= \frac{74}{100}\{1040\}\times\frac{2,00,000}{10^6}kg = 153.93kg$$

$$\text{Total req. of Soda} = \frac{106}{100}\{400+300-600-10+50+50\}\times\frac{2\times10^5}{10^6}kg$$

$$= \frac{106}{100}\{190\}\times\frac{2\times10^5}{10^6}kg = 40.28kg$$

5. A water sample on analysis gave the following data:-
$CaCl_2$= 11.1ppm, $CaSO_4$= 70ppm, $Ca(HCO_3)_2$= 300 ppm, $MgCl_2$= 9.5 ppm NaCl= 140 ppm, $MgSO_4$=48 ppm, $Mg(HCO_3)_2$=170 ppm.Calculate amount of lime(90% pure) and soda(94% pure) required to soften 10,000 liters of water using $NaAlO_2$ as coagulant at the rate of 16.4 ppm. Assume the concentration of OH^- and CO_3^{2-} in treated water as 5 and 15 ppm respectively. Justify your answer with chemical equations.

Ion-exchange process of water softening

Ion-exchange process is the reversible exchange of ions between a liquid phase and a solid phase. Materials capable to of exchange of cations are called cation exchangers and those which are capable of exchanging anions are called anion exchangers. Ion exchangers commonly used in water treatment include:
(i)   Natural and synthetic zeolites
(ii)   Carbonaceous ion exchangers
(iii)   Synthetic resins

Zeolite or permutit process

The zeolites are hydrated sodium alumino silicate. The general chemical formula of zeolite is

$Na_2O \cdot Al_2O_3 \cdot xSiO_2 \cdot yH_2O$ where x varies 2 to 10 and y varies 2 to 6. Zeolites are capable of exchanging reversibly sodium ions for hardness producing ions in water. They are of two types

(a)   Natural zeolites are nonporous. For example, natrolite, $Na_2O \cdot Al_2O_3 \cdot 4SiO_2 \cdot 2H_2O$. This generally derived from green sands by washing, heating and treatment with NaOH.

(b)   Synthetic zeolites are porous and posses gel structure. They are prepared by heating together china clay, feldspar and soda ash. Such zeolite possess higher exchange capacity than natural one

Process of softening

When hard water passes through a bed of active granular sodium zeolite, Ca^{2+} and Mg^{2+} from liquid phase get exchange with sodium. The various reactions taking place during this process are:

$$Ca(HCO_3)_2 + Na_2Z \rightleftharpoons CaZ + 2NaHCO_3$$
$$CaSO_4 + Na_2Z \rightleftharpoons CaZ + Na_2SO_4$$
$$MgCl_2 + Na_2Z \rightleftharpoons MgZ + 2NaCl$$

where Na_2Z is the zeolite. Small quantities of iron and manganese also get removed along with Ca^{2+} and Mg^{2+}.

Regeneration

When the zeolite bed is saturated with Ca^{2+} and Mg^{2+} (exhausted) it ceases to soften water. The exhausted zeolite can be regenerated and reused. This can be done by treating zeolite bed with concentrated (10 %) brine (NaCl) solution.

$$CaZ (or\ MgZ) + 2NaCl \rightleftharpoons Na_2Z + CaCl_2 (or\ MgCl_2)$$

Advantages

(i)   It removes hardness almost completely (hardness about 100%)

(ii)   The equipment used here is compact.

(iii)   It automatically adjusts itself to water of varied hardness.

(iv)   It operates under pressure.

Limitation

(i)   Water containing excess of acidity or alkalinity may attack the zeolite.

(ii)   Acid radicals are not removed by this process because it produces $NaHCO_3$ which breakdown in boiler and form NaOH. The presence of NaOH leads to a caustic embrittlement in the boiler.

$$NaHCO_3 \longrightarrow NaOH + CO_2$$

(iii)   The byproduct CO_2 in above reaction renders water acidic corrosive.

(iv) Fe^{2+} and Mn^{2+} when passed through the zeolite bed are converted into their respective zeolites which cannot be regenerated.

$$Fe^{2+} + Na_2Z \longrightarrow FeZ + 2Na^+ \quad \text{(Irreversible)}$$
$$Mn^{2+} + Na_2Z \longrightarrow MnZ + 2Na^+ \quad \text{(Irreversible)}$$

Comparison between zeolite and lime-soda process

Zeolite process	Lime-soda process
1. Water of 10-15 ppm hardness is obtained	Water of 15-50 ppm hardness is obtained
2. Water contains large amount of sodium salt	Water contain less amount of sodium salt
3. It can operate under pressure	It cannot operate under pressure
4. It cannot be used in acidic or basic water	There are no such limitation
5. Cost of plant is higher	Cost of plant is lower
6. Operation expenses are	Operation expenses are

lower	higher

Zeolite calculation

1. The relation for determine hardness if volume of water and the amount of NaCl are given for the complete softening of water.

'V' Lit of hard water $= v$ Lit of NaCl

$= v \times c$ g of NaCl,   (where c is the strength of NaCl)

$= v \times c \times \dfrac{100}{(2 \times 58.5)}$ g of $CaCO_3$ eq.   (where 58.5 is the Mol. wt of NaCl)

'V' Lit of hard water $= v \times c \times \dfrac{100}{117} \times 1000$ mg of $CaCO_3$ eq.

1 Lit of hard water $= v \times c \times \dfrac{100}{117} \times \dfrac{1000}{V}$ mg/L of $CaCO_3$ eq.

$\therefore$ Hardness of water $(y) = v \times c \times \dfrac{100}{117} \times \dfrac{1000}{V}$ mg/L

2. The relation for determine total volume if hardness of water and the amount of NaCl are given for the complete softening of water.

Volume of water $(V) = v \times c \times \dfrac{100}{117} \times \dfrac{1000}{y}$ L   (where y is the hardness of water)

3. If NaCl is given in percentage

$\left[c = w\% = \dfrac{w\,g}{100\,ml} = \dfrac{10w\,g}{1000\,ml} = 10w\ g/L \right]$

Hardness of water $(y) = v \times 10w \times \dfrac{100}{117} \times \dfrac{1000}{V}$ mg/L

Volume of water $(V) = v \times 10w \times \dfrac{100}{117} \times \dfrac{1000}{y}$ L

4. Calculation of weight (W) of NaCl

Wt. of NaCl (W) $= V \times \dfrac{117}{100} \times \dfrac{y}{1000}$ g

Examples on zeolite calculation

1. The hardness of 50, 000 L of a sample of water was removed by passing it through a zeolite softener. The softener then required 200 L of NaCl solution (brine) containing 150 g/L of NaCl for regeneration. Calculate the hardness of the water sample.

    Hint: Use   formula:

    $$\text{Hardness of water } (y) = v \times c \times \frac{100}{117} \times \frac{1000}{V} \text{ mg /L}$$

2. An exhausted Zeolite softener was regenerated completely by passing 400 L of brine solution containing 3.5% NaCl. How many litre of hard water sample having hardness 560 ppm can be softened using this softener?

    Hint: Use   formula:

    $$\text{Volume of water} (V) = v \times 10 \, w \times \frac{100}{117} \times \frac{1000}{y} L$$

3. The hardness of 20,000 litres of a water sample was completely removed by passing it through a zeolite softener. The zeolite softener required 220 litres of NaCl solution for regeneration. If the water has hardness equivalent to 171 ppm of $CaCO_3$, calculate the concentration of the NaCl solution used.

    Hint: Use formula:
    $$c = \frac{y}{v} \times \frac{117}{100} \times \frac{V}{1000} \text{ g/L}$$

4. An exhausted zeolite softener was regenerated by passing 117 L of NaCl solution, having strength of 100 g/L of NaCl. If the hardness of water is 600 ppm. Calculate the

total volume of water that is softened by this softener.

$$V = v \times c \times \frac{100}{117} \times \frac{1000}{y} L$$

Hint: Use formula:

Ion-Exchange or deionization or demineralization process

Ion-exchange process is the reversible exchange of ions between a mobile liquid phase and stationary solid phase. This process essentially contains two columns one filled with cation exchange resin and other is anion exchange resin. Resin consists of an insoluble polymeric matrix. It has fixed charge groups with counter mobile ions of opposite charge. Mobile counter ions are capable to exchange with ions of same charge from liquid phase. The formula of typical cationic and anionic resin based on styrene and divinylbenzene:

The sulphonated form of cation exchange resin contains H^+ exchangeable ion, represented as RH^+ and anion exchange resin has OH^- exchangeable ion.

Process

(i)   The hard water is allowed passed through cation exchange column, which removes cations and release equivalent amount of H^+.

$$2R^-H^+ + Ca^{2+} \rightleftharpoons R_2Ca + 2H^+$$

(ii)   The hard now is passed through anion exchange column, which removes the anions and equivalent amount of OH^- are released.

$$R^+OH^- + Cl^- \rightleftharpoons RCl + OH^-$$

$$2R^+OH^- + SO_4^{2-} \rightleftharpoons R_2SO_4 + 2OH^-$$

(iii)   The H^+ from cation exchange column and OH^- that from anion exchange column combine and form water molecule.

$$H^+ + OH^- \longrightarrow H_2O$$

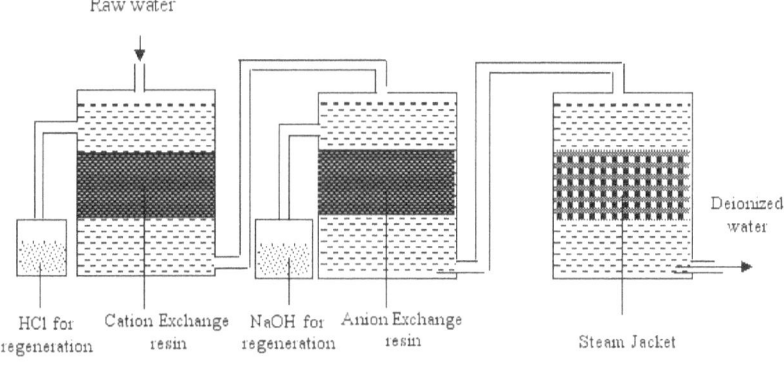

(v)   Thus the water coming out from the exchanger is free from cations as well as anions. Ion free water is known as deionized or demineralised water

Regeneration

When capacities of cation and anion exchangers to exchange H^+ and OH^- ions respectively are lost, they are said to be exhausted. The exhausted cation exchange resin is regenerated by passing a solution of dil. HCl:

$$R_2Ca + 2HCl \longrightarrow R^-H^+ + CaCl_2$$

The exhausted anion exchange resin is regenerated by passing a solution of dil. NaOH:

$$RCl + NaOH \longrightarrow R^+OH^- + NaCl$$
$$R_2SO_4 + 2NaOH \longrightarrow 2R^+OH^- + Na_2SO_4$$

Advantages of demineralization

1. Highly acidic or alkaline water can be softened by this process.
2. It produces water of very low hardness (2-5 ppm).
3. The water produced by this method can be used in high pressure boiler.

Comparison of lime-soda, zeolite and deionization

Lime-soda	zeolite	deionization
Dissolved salts are converted into insoluble ppt by treating them with lime and soda. Water of 15-50 ppm hardness is obtained. Treated water contained large amount of sludge & ppt.	Reversible exchange of hardness causing cations with Na ions of zeolite. Water of 10-15 ppm hardness is obtained. Treated water contained large amount of sodium salt. More acidic or alkaline water cannot be softened.	Reversible exchange of cations and anions with exchangeable cations and anions of resin. Water of 2-5 ppm hardness is obtained. Treated water contained negligible amount of sludge. Acidic or basic water can be softened.

Acidic or basic water can be softened.		

www.ingramcontent.com/pod-product-compliance
Lightning Source LLC
Chambersburg PA
CBHW030551220526
45463CB00007B/3063